Maldon Shipwrights

David Patient
James Dodds

Jardine Press
2020

Dedicated to
Barry Pearce

Jardine Press Ltd 2020
ISBN 978-0-9934779-8-0
Text © David Patient
Images © James Dodds

Introduction

Maldon on the river Blackwater in Essex has a proud tradition of building and maintaining the shipping of the Thames. In particlular the Thames Barge, a unique craft in rig and construction for carrying cargos to and from the great London docks and the shallow waters and creeks of the Thames estuary.

In 1972 aged 15 I started working at the barge yard of Walter Cook and Son with the boatbuilder Alf Last. This was followed by a year at shipbuilding school in Southampton. On return I began working alongside David Patient. He had an art college background, an amateur boatbuilding career and had resigned from teaching art in Maldon. Cook's was then under the management of Barry Pearce & Gordon Swift. Baden Dedman was our foreman and he worked with Bill Claydon. The main work was converting wooden motor barges back from freight to sail and carrying people. The skills and tools used in this small yard would have been very familiar to Noah and his sons!

Having completed my four year apprenticeship I went off to Chelsea School of Art and then the Royal College of Art. These formative years in boatyards became the inspiration for my artistic career.

David went on to build a replica of a Gravesend shrimper 'Marigold' at Cook's and then to run his own shipyard higher up the river at Fullbridge. The second half of this book shows the work I was involved with when returning after art school to work part time for David in the early '90s.

'Maldon Shipwrights' celebrates our frendship and time spent working together. This book contains my linocut memories and is accompanied by David's captions.

In retirement David has been able to finish his research into the history of the pioneering Maldon bargebuilder John Howard (1849-1915), which is also published by the Jardine Press in 2020.

James Dodds

Maldon Shipwrights

The scene is set from memory, principally illustrating a difficult operation involving all those employed at Cook's, in this instance in January 1976, including Jamie and me. A new bow-rail for the barge *Ethel Ada* is being bent over a former. This process involved cutting out the rail to the required shape on the flat from a three inch board of oak. The name of the barge is carved in at this stage for working convenience and the former is built up to represent the bow shape of the barge.

Jamie remembers being given the task of pumping up the oil for the tug *Chrianie*'s engine to provide steam. This is being piped under pressure into a steam chest to soften the oak plank within. After having been given the required steaming of one hour for every inch of the thickness of oak, the plank is removed for bending.

The men are seen, with 'Armstrong Patent' and cramps, bending the timber as quickly as they can before it cools. The shipwright, unnecessarily standing on top, is distracted by a young girl walking by the shed.

Jamie depicts barges on the yard which were sold out of trade to be repaired and re-rigged for charter work locally; this was Cook's main work at the time. The lad to the right could be Jamie who initially helped the boatbuilder Alf Last primarily building 14ft barge boats. As a boy apprentice himself Alf was given a three month trial by Walter Cook to see if he passed muster but Alf never heard another word about it. Retiring from the firm after 57 years service he remarked that he still didn't know whether he was taken on.

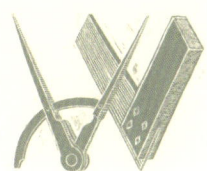

Floodtide

'Tide work' involves working below the waterline on the hull at low water. A barge has been previously floated onto a set of blocks supported off the foreshore at a height to just enable a shipwright to get underneath if necessary. A section of side planking has been renewed and the last of the spikes is being driven with the tide approaching the shipwright's thigh boots. He wields his maul with force and accuracy with no time to lose. The suspended bucket contains his brace and bit to drill the holes and 6 inch spikes which are heavy duty 'nails' used for fastening the planking.

Also prominent and nicely drawn is the barge's starboard leeboard; visitors to the yard and quay are often baffled as to their use. If the locals are in a good mood they will explain that they are a substitute for a keel on a flat-bottomed sailing barge. Let down below the bottom on the leeward side, they 'grip' the water to prevent leeward drift and encourage forward movement.

This print shows Jamie's interest in portraiture and the capturing of a process experienced during his time as a shipwright. In a series of linocuts, whist maintaining his story telling, he went on to affect a more scenic approach with flowing compositions.

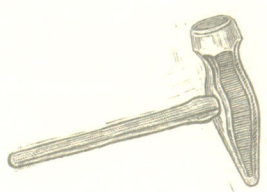

Springtide

Jamie continues with his early memories of his time at Cook's. He depicts himself learning how to scull a barge's boat. This involves the use of one oar set in a notch in the transom and a figure of eight movement creates the pressure to propel the boat forward.

The sailing barge is seen on the yard's set of blocks for work to be done below the waterline at low tide and she is being rigged out for the sailing season. The mast and spars on a barge are designed to be lowered for going under bridges and to facilitate maintenance, and in this instance the mainsail is being bent on.

To the right Alf Last, the company's boatbuilder and Jamie's mentor, has come out of his shed. Alf is pictured wearing clogs instead of his usual footwear; this is Jamie reverting to symbolism, a feature of this early work. Jamie was reminded of a group of young Dutchmen that had descended on a sailing barge at Maldon's quay and stayed to make an impression.

Perhaps for the first time in his linocuts Jamie has distorted the image of a barge. As far as the nautical minded among us are concerned, this liberty is only made acceptable because of Jamie's intimate knowledge of his subject. The exaggerated lines have allowed him to create a flowing composition combined with a recognisable representation.

Mast Making

Two shipwrights are working together to produce a mast. One is swinging the adze above his head to remove large areas of timber with force, the other is taking off ' beeswings'. Jamie from the start of his apprenticeship had to quickly master this skill, handed down from our Viking ancestors. For a barges' topmast for instance, larch or Douglas fir was obtained from a specialist timber yard. Often at Cook's the 70ft trees arrived in the round which amounted to extra labour having to square them off first. Timber mills were able to cut a spar square and even to the required taper.

A round tree is depicted with the surplus being adzed off on one side, delineated by a chalk line. Rough saw cuts, just short of the line, are made regularly to allow sections to be cut away more easily.

The face is regularly checked with a level to end up perfectly plumb with no twist. The worst is done and the surface is turned uppermost to continue shaping, guided this time by the faithful try square applied to the working face. When the spar is square the process for rounding is facilitated by a spar gauge run along the length of the timber marking out the eight sides. Draw knives and hand planes come into use at this stage and the reduction to sixteen sides and the final rounding is usually judged by eye.

Marigold

The *Marigold* is a copy of the clinker-built Gravesend shrimper *Lilian* built at Gravesend in 1869 and these vessels became commonly known as bawleys. The *Lilian* was partially rebuilt by me over the years and as a young shipwright only experienced with repairs, ashamedly abandoned her to be able to build a new vessel. She was sponsored by Barry Pearce at Cook's and Jamie and I took patterns from the *Lilian* and he, college educated, drafted the lines. I built the vessel between barge work and Jamie has represented the *Marigold* in the summer 1979 on the completion of the hull. Cook's 'tearoom', a place to gather for passing fisherman and bargemen, is featured along with one of Alf's standard 14ft barge boats still in demand for barges and to be rigged out for sailing. Nicely featured to the left is the leg-vice attached to our long bench and a very clever use of perspective of the view of the river if we could have seen out of the windows.

It is not an exaggeration to suggest that Jamie has followed in the footsteps of the marine artist E W Cook (1811-1880) who was prolific from the age of 17. Without his painstakingly accurate observations we would today be unaware of the exact design of certain types of 18th century sailing craft. Jamie has in turn appreciated, researched, drawn, painted and printed a variety of working craft which are no long with us. Further he has also taken a particular interest in vessels that have been saved or built new on traditional lines; this print of the *Marigold* is a fine example.

Early Morning Tide

This linocut celebrates Walter Cook & Son's centenary in 1994. The barge yard site was originally established circa 1855 and was run by the shipwrights Handley and Finch who were occupied with local repair work. The land was leased from the Maldon Council (as it is today) and taken over by Walter Cook and Arthur Woodward to build the sailing barge *Dawn* launched in 1897. They both had previously worked for John Howard at his Maldon barge yard. Cook and Woodward's livelihood was sustained with repairs but they built two more barges, the *Lord Roberts* and *British Empire*. Woodward left the partnership and Cook concentrated on boatbuilding until after the Great War when he was joined by his son Cliff and took on barge repairs once again.

The yard was saved in 1970 by Barry Pearce and Gordon Swift before the Maldon Council could turn it into a car park. Jamie has pictured the new shed and to the right the barge *George Smeed* under extensive rebuild on 'Noddy' Cardy's pontoon. Apart from its use for the replanking of the *Gladys*, seen later, this facility provided full time access as opposed to the restrictions imposed by the tides with the barge *Beric* on the traditional blocks, seen to the left.

We got half-way through converting the *Beric* back to sail when the owners wanted her to carry wheat for one more season. For her to again function as a motor barge this involved rebuilding the high coamings and wheel house we had removed.

In the centre can be seen the shipwright Ian Danskin on the foreshore 'puggling'. This involved the ancient method of disturbing the mud on the ebb tide that is continually deposited on the flood. A flat piece of timber is secured at right angles to the end of a long pole, like a wooden hoe, which is worked back and forth creating a wash unsettling the silt.

As far as 'art' is concerned Jamie has extensively used his 'distortion' tactics to provide a very flowing image, appropriate for an evocative maritime scene.

Building The Eleanor Mary

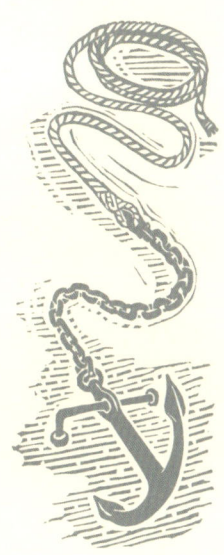

I was building boats in 'Howard's' old blacksmith shed at the bottom of North Street after leaving Cook's and went on to establish a yard at Fullbridge to build the yacht *Eleanor Mary*. She was a commission from John Lamb of Maldon from a design by the late Alan Hill of Burnham on Crouch. A scaffold pole and canvas shed was built for the purpose.

The yacht is being planked up and Jamie obviously took an interest during its construction by producing a beautiful pencil drawing and this linocut. We posed for a photograph by Jamie in the tradition of shipwrights standing in front of the vessel they had built holding their tools of trade. As a gesture to modern life 'Dutch' Pete, to the left, holds a spade and I, to the right, hold a chainsaw.

Shipwright Peter Graham is pictured in the middle; we are all immediately identifiable. This requires considerable skill from Jamie bearing in mind that likeness is accomplished by cutting a small section of lino. Barry Pearce, Jamie's boss at Cook's, commented 'That's bloody clever that is'.

The *Eleanor Mary* proved to be a very fast cruising yacht; getting ahead of the rest of the fleet in a local Old Gaffer's Race.

Shipwrights Yard

Here Jamie has represented my barge yard for the first time, established at Fullbridge after building the yacht *Eleanor Mary*. He had for some time been happy to come back to shipwrighting especially when the art world was quiet. Indeed he has depicted himself caulking the decks of the Baltic Trader *Queen Galadriel* and the fumes from the pitch boiler are a feature. My most extensive enterprise lies alongside; the building of a complete new 'top' on the *Ardwina*. A tent was built over the barge but with Jamie's ex-ray vision all the new deck beams are visible. The yard's 35 ton Preistman crane is represented and in lieu of his interest in symposium Jamie has included items that only mean something to him and me. He helped with cutting out the timber for the deck planking and lodging knees and in the background the timber is depicted stacked and covered for seasoning. I am just entering my shed talking on a new remote phone. The compact and circular composition was to be repeated in subsequent linocuts commissioned to represent the yard's work and sent out as Christmas cards. Very many friends still have the cards on display; a testament to how well Jamie's early print work was appreciated.

Caulking

For carvel-built working craft caulking primarily involved the driving of oakum into hull and deck plank seams that are bevelled for the purpose and sealed with hot pitch. Oakum is rolled (spun) by hand into a strand of a thickness dictated by the width of the plank seams.

The caulker, illustrated, holds a wooden caulking mallet especially designed to withstand constant pounding and he is about to strike a caulking iron. A narrow shaped iron is used initially to introduce the stands of oakum into the seam, followed by the use of a blunter iron to drive the oakum deep enough to allow a thickness of pitch. The pitch was heated in a boiler close at hand and transferred to a ladle for pouring into the seam; a messy job in a high wind.

Jamie was taught the art of caulking at Cook's; not an inconsiderable skill. We have come across permanently damaged planking caused by many a so-called 'caulker'. In times past caulking was a separate occupation in the shipyards but in the small yards it was part of the shipwright's work.

Caulking on barges at Cook's would only involve deck work as a barge is built without a caulked hull. Their construction involved 'set work', where the planks are set down tight against each other with a compound in between made up of tar, pitch and animal hair.

Gladys

The barge *Gladys* worked under sail and motor for millers at Ipswich and was re-rigged at Cook's under the same ownership for hospitality work. We took over her maintenance and she is seen having been extensively replanked on a pontoon at Heybridge Basin. The pontoon was supplied by 'Noddy' Cardy, waterman, tug owner and marine contractor in Maldon who was responsible for this specialist operation. The pontoon is sunk, the barge is floated into position and held there until she grounds on a set of blocks. On low water the pontoon is pumped dry allowing it to lift its cargo on the next tide.

Jamie was on hand to help out with the rebuild and continued to capture nautical scenes for his art work. Subsequently he has pictured Peter Graham, Adrian Riva and myself at work and the mate leaning on the rudder with the skipper, the late Terry Everett, by the steam box. The barge is surrounded with rough seas and bellowing clouds which stands in contrast to the attention to detail Jamie has taken over the image of the barge and pontoon.

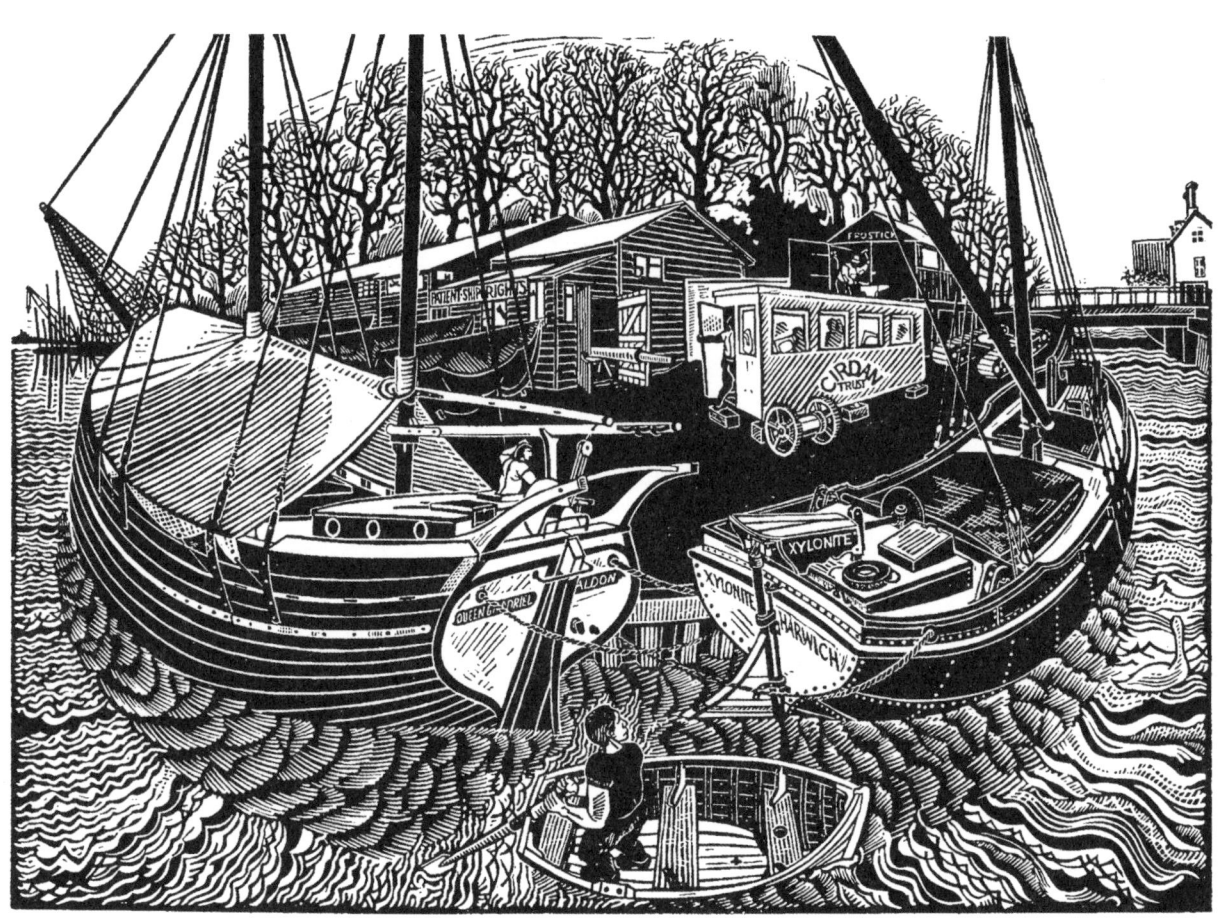

Winter Refit

The yard is pictured when it was well established and we were working hard for the Cirdan Sailing Trust having taken charge of their new acquisition, the steel barge *Xylonite*. I was then also their landlord with their office on the quay. The barge is seen unrigged for maintenance and her windlass is ashore for Jamie to repair. Other small items have to be looked for; over the back is the blacksmith Tony Frostick at this anvil. He was responsible for most of the well formed steelwork the yard needed. The old work shop has its door open to enable us to make a new topmast for the Cirdan's *Queen Galadriel* that was much longer than the shed. The bottom half of my Priestman crane is seen to the right; the top has been lifted off its tracks for an extensive re-build. To the extreme right Jamie has included the bridge over the River Chelmer and the Welcome Sailor public house. At this time the land behind the yard had not been developed and the trees depicted make a splendid backdrop. In the foreground the *Xylonite's* boat is being sculled; this skill was a must for all bargemen as it was more of a convenient means of propulsion than rowing. Jamie has drawn heavily on his previous composition of the yard but triumphantly succeeded in incapsulating all that was going on that winter and producing an independent art work into the bargain.

New Topmast

My old Fullbridge work shop was a lean-to attached to the back of a 1st World War Nissen Hut.

Jamie and Pompy (David Abadom) are making a new topmast for the Baltic Trader Queen Galadriel. The 'big' bandsaw has been captured along with the bottom part of my legs disappearing up the office steps. As the spar was too long for the work shop, a hole was cut in the end door for it to poke through and allow us to lock up for the night.

This old shed was burnt down in 1998; the fire destroying a newly built 15 ft 'winkle brig' and a much prized collection of hand tools and machines.

We ended up in a 40 foot steel container until the yard was resurrected by a local developer who built me a brand new workshop.

Launching Hartlepool Renaissance

The Reverent Bill Broad was the founder of the Cirdan Sailing Trust which helps underprivileged children within a sail training situation. He wanted to develop the charity 'up north', hence the acquisition of a yacht that was renamed the *Hartlepool Renaissance*. We took charge of the yacht for a complete overhaul at Fullbridge. Bill was good with people and I still remember an instance of this thoughtfulness. He asked if he could be entirely responsible for the ordering and installation of the yacht's new electronics. On her maiden voyage around Britain I got a call from the yacht. Steeling myself for the usual owner's rebuff that some disaster was my fault, Bill apologised for his involvement as all the electronics had failed and that everything else was holding up well.

A heavy lift road crane was ordered for the launch; the yacht is about to be swung into the river. In true 'renaissance' tradition Jamie has included his patrons. Featured in the group of onlookers are my wife Jess and our two young children and I am seen beside the crane cab unnecessarily instructing the driver.

Steaming New Planks on the Scone

The sailing barge *Scone* was built in 1919 and after her trading days as a motor barge ended, she was converted back to sail. The barge had 'come of its wants', an old expression coined by the local bargebuilder Cliff Cook. She was at Fullbridge for extensive repairs to her bow planks and frames. Scaffolding had to be set up to facilitate the work afloat. The barge's side was planked in two layers of an inch and a half planking with tar and felt between, and never caulked. The scene shows Jamie and me pulling a hot plank out of the steam chest with the fire under the boiler going well. The young shipwrights Peter Graham and Adrian Riva are bending another plank round a makeshift former. Up the barge's rigging can be seen the bargeman Geoff Harris who is steadily ascending, bending on new rattlings to the shrouds which on the starboard side act as a ladder to reach the topsail.

Jamie's fascination with the complexity of the industrial architecture over the river, has resulted in a wonderful composition. The mill was once supplied with grain by a fleet of sailing and motor barges, including the *Scone*.

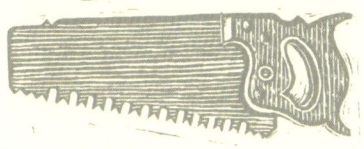

Fullbridge Shipyard Maldon

This linocut was a gift from Jamie on my forthcoming retirement. He has included in it a number of images that relate to the life and times of the yard at Fullbridge.

The bottom of the new housing development became our new boatshop and The Cirdan Sailing Trust took over the upper story for their office. Tony Frostick also had a new smithy off the road.

The old Gravesend shrimper *Lilian* is seen in strops being readied to be taken away for a final rebuild by a young shipwright at Pin Mill. The *Queen Galadriel* projects her bowsprit into the scene from within a dry dock moored above the yard available for all the underwater work. This was a new addition to 'Noddy' Cardy's fleet, replacing his pontoon. To the left is the bow of the oyster smack *Varuna* in frame; an infill job over the years and launched as a completely new vessel before retirement. In front of the crane is our own boat, a Dutch Zeeschouw. The barge *Violet* to the left is purely a reminder of my research into John Howard the Maldon barge builder who was responsible for her build in 1889. The *Spitfire*, top right, also reflects a lifelong interest in vintage aircraft.

For this personal linocut Jamie has used a more linear approach rather his usual fluid technique.

Man with an Oar

It was a common expression in the boatyard that when you got fed up with everything to do with boats you would say 'I'm going to put an oar over my shoulder and walk inland until somebody asks me what it is. I shall plant it in the ground and that's where I shall stay'. Only much later did I discover that this comes from *Ulysses* when Odysseus returns from his voyage.